WITH APPRECIATION TO:

Through the years I have worked with the most skilled "masters of fusion splice". Usually at a 06:30 training session...there are..."a few splicers"! Oftentimes seminars for "end face cleaning" were confused with "fusion splice prep". The two cleaning processes are different. Lack of awareness of cross-contamination may be a result of "non-applications-specific" training. This book adds to the knowledge base: it's intention is to "Future Proof".

This work is written for you: Sumitomo® or AFL®? Corning® or Fitel®? A "house brand" or hand-me-down? The information is important. A fusion splicer is not a 'casual expense'. In researching the topic I learned a little about what some might consider 'niche devices': Aurora Optics®, produces Made-in-USA machines with intriguing technical features.

If you have a fusion splice machine, likely someone 'squeezed a budget' to make sure your work is top notch. Make sure you protect this investment. A low-loss, first time splice was once possible for reasons that were somewhere between "magic and skill"! Contemporary fusion splice machines come with many features: be sure you understand not only the equipment, but also the environment where you will use it! Application of an 'explosion-proof' machine may seem narrow, until time a splice is made in a mine, fuel depot or sewer with elevated methane levels. As fiber optics replace copper, there are many venues.

For this work I want to acknowledge Darrin Newman of Fitel®. His knowledge of fusion splice and the intricacies of these instruments helped make this work important to all. Perhaps more so, Laurence Wesson of Aurora Optics® provided invaluable insights into applications specific topics surrounding explosion proof fusion splicing and measured differences between various splice prep solvents in which I found no small interest. I am appreciative to both for suggestions and edits.

Through the years I have been on the road with manufacturer's reps, whom I prefer to call "sales engineers". Among these highly talented "masters of fusion splice" are Paul Looney in the Northeast-USA, Chuck Mason in the Midwest-USA and Jerry Leslie in the Southeast-USA. There are many others.

In recent times I have been honored to associate with former "competitors and foes"! These works are written in a 'non-commercial' way and understandably troublesome (perhaps) to some commercial interests. "Science" is both black and white and also possible in gray scale! While there may be commercial disagreement, I believe it is essential to understand matters of science as related to precision cleaning and the future of our Industry. Individuals and manufacturers noted herein may or may not agree with the thesis of this work, or, any in the series.

As described in "How To Precision Clean a Fiber Optic Connection", there are myriad physical venues and ambient sources of debris and contamination. Discussed here, another possible contamination is at the splicer...which is open and unprotected...and subject to visual and microscopic contamination that can influence the end result. The marginal condition of some of these devices (some well-maintained and others not-so-much) inspired this work.

Some years ago, with another mentor, Ken Putnam, several hundred technicians were trained in a garage-to-garage "Tour de FTTp" in California. The session always started with the question: *"who is cleaning a fiber optic connection?"*. The answer was surprising: nearly 70% were prepping for fusion splice and the others were cleaning the "end face". Nearly everyone thought (at the time and often still do) the two procedures are one in the same with interchangeable products. *Fusion splice prep is not end face cleaning*. The two are different applications with different tools and very different needs.

Finally, in the first place, I want to thank my wife, Lanet, and others in the family who endured the times during all hours of the night, perhaps on vacation, I would be tapping keys on the computer as they calmly closed the bedroom door in places all over this wonderful world!

1

WHAT'S INSIDE:

INTRODUCTION: Why is this book is written?

Recently I turned left onto a main road near my residence. There was a splice trailer and a backhoe: *never a good combination*! There was a cut and the technicians were struggling to restore service: tomorrow is SuperBowl® Sunday!

Whether a splice-on connector, an emergency repair, or standard network operation, fusion splicing has evolved rapidly. Second Generation craftsmen tell how First Generation "old timers" would weld glass fibers with a gas torch. It was not so long ago that fusion splice machines were huge beasts; now they are pocket size and highly adept at low loss splices. With the current generation of fusion splice equipment (as with so many things in fiber optics) what was once thought theoretical is in practice today.   There are 'explosion proof' and 'mini-splicers' that can be used (inverted) in tight places such as the fuselage of a high performance military aircraft...or behind a cabinet in a 'wire maze'!

Whatever tomorrow brings for fiber optic deployments, one thing is sure: the work will be done hanging in a bucket or deep in a trench, comfortable in a lab, or, in misery in a sand storm or under a dripping tarp … somewhere! The work environment is infinite and so the potential for contamination. This contamination may also come from the products used to precision clean. These are commonly called "residues" and they can arise from: 1.) inadequate cleaning materials or, 2.) improper technique.

A contemporary fusion splicer is a technical marvel of fiber optics that requires critical but 'simple' maintenance. How often? That depends on ambient conditions and the total splice cycle count. As a carry-on work to *"How To Precision Clean All Fiber Optic Connections"* and *"A Study of Precision Cleaning Processes: What Works and What Does Not"*, this tutorial explains how to and why "exceeding standards" is integral to "future proof" fiber optic deployments.

As is the case with the connection end face and optical equipment, contamination at various points in the fusion splicer can add 'distortion' and add to the 'loss budget'. As with any fine piece of equipment, every fusion splicer should be regularly cleaned to assure not only that splice, but also the next one will not be a source of loss.   If you are using a fusion splicer you are entrusted with a costly instrument you purchased, leased or was extracted (possibly at great effort) from a 'budget'.   These simple and effective maintenance procedures will assure low-loss, first-time results.

It is likely that your manufacturer has a product and possibly even instructions on how to maintain your investment: always defer to the manufacturer! This work is intended to "exceed standards" by enhancing existing techniques and suggesting areas of possible concern and process improvement. At least follow the manufacturer's recommendations, and exceed, never compromise maintenance intervals.

Precision cleaning 'just about anything fiber optic' is in an infancy stage: the goal is to "future-proof".

What is what meant by cross-contamination?
What is 'cosmetic cleaning'?
What is 'precision cleaning'?

- A dark smudge
  on a piece of white paper …
- A finger print on clean glass…

"CROSS-CON ·TAM ·I ·NA ·TION" *

noun: cross-contamination
the process by dust. Fluids or other micro particles are
unintentionally transferred from one substance or object to
another, with harmful effect.

The <u>transfer</u> of a <u>contaminant</u> from one source to another.
* February 1st, 2015 Google Search in Wictionary™

- There are myriad examples. In the instance of precision cleaning test equipment and fusion splicers dust or oils on the outer cases and surfaces can be transferred to the mechanical surfaces.

- Proper maintenance of fusion splicers assures long life of this expensive device. Whether your instrument is new or vintage, surfaces such as electrodes, lenses, mirrors and fiber holders must be properly maintained.

- Likely, the producer of the equipment has suggestions and always read these carefully.

- These components should be visually inspected before each splicing session. As would a 'good scout', the instrument should be 'cased up and stored" ready for the next time. "Cosmetic cleaning" assures "precision components" are not effected.

- If you are working in a harsh environment or clean room, the job doesn't end with the last splice! The job is finished when the equipment is stored in better condition than when you started the session.

- Please, when you are "given" one of these instruments, perhaps extracted from a limited budget, maintain it.

Why is loss in a fusion splice important?

As a fiber optic installation is designed, installed, or, serviced the nature of light demands certain factors be considered to assure the light will make the trip from transmission to reception!

The first is the reality that there is an intrinsic loss of power because of the nature of light. Others include loss at connectors, couplers, splitters and...loss as the power passes through splices. The designer calculates these factors on paper; a quality fiber optic power meter will assure your installation is within "loss budget".

The diagram below, depicts three types of loss: 1.) Fiber loss along the transmission length (transmitter to receiver), 2.) connector loss (3 connectors) at the connection, 3.) splice loss (1 splice) along the run. *These three units combine to establish the "loss budget".*

Since a fusion splicer is a computer-controlled device that plasma welds either single or multiple fibers simultaneously, maintenance of the moving parts and components on the instrument can and will influence the end result.

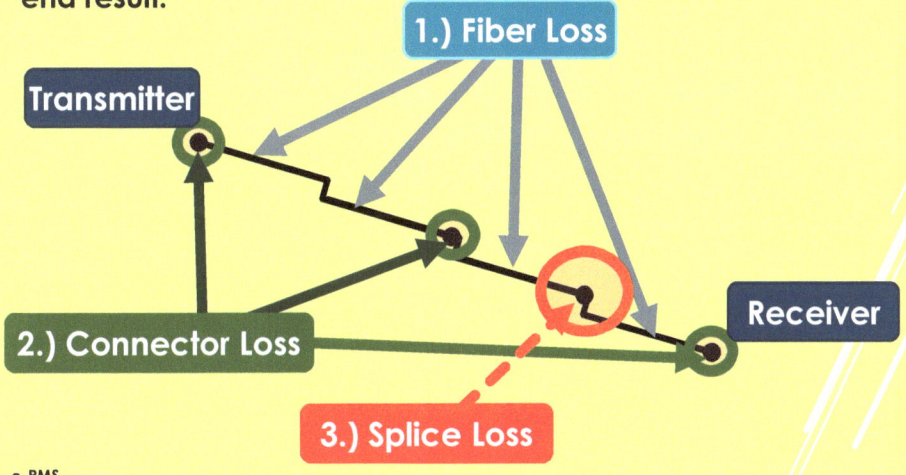

c-RMS

What is a fusion splice?

Fusion splicing is the act of joining two or more fiber optic fibers using heat. The source of heat is usually an electric arc.

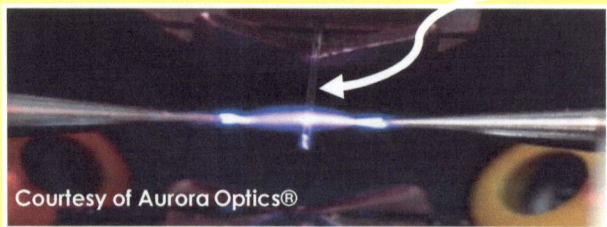

Courtesy of Aurora Optics®

This process is accomplished by fusing fibers in such a way that light is not scattered or reflected by the splice. This procedure is performed by sophisticated instruments called fusion splicers.

There are various fusion splicer types. One type aligns the fiber to be spliced by the 'cladding" (2). The other aligns the splice keying off the 'core" (1). A "core" on single mode fiber is 9 microns and the "cladding" is 125 microns: a multi-mode core can be 50, 62.5 or 100 with a 125 or 140 micron cladding. The x-y-z alignment systems on fusion splicers calculates these dimensions and typically make an effective splice in less than 60 seconds!

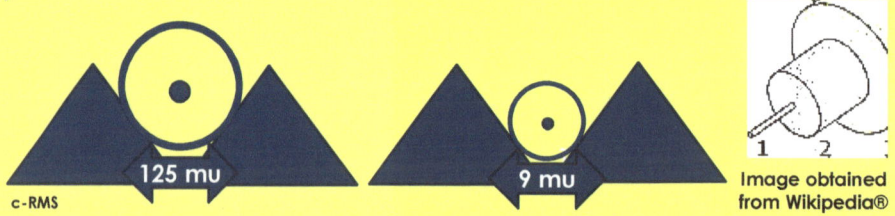

c-RMS

Image obtained from Wikipedia®

To accomplish this amazing technological feat all fusion splicers have an array of motors, alignment devices and precision holders. Some of these are accessible and can be precision cleaned. Others are sealed and not serviceable. It is important you read the instructions for your specific machine to determine the components such: 1.) V-Grooves, 2.) Fiber Holders, 3.) Mirrors, 4.) Lenses, 5.) LED surfaces and any other operational components that may require precision cleaning as an integral part of normal use. It is also important to "cosmetically clean" the machine.

There is also "abnormal use": installation in inclement weather or "unusual environment" such as a coal mine or dusty office back room! Debris on the outside of the machine can contaminate the internal precision components: don't under-estimate the need to cosmetic clean!

There are proponents of "v-groove", "cladding". "core" alignment as well as ribbon splicing machines: challenge the supplier: *"Which fits my need"!*

"Next Generation" Fiber Optic Fusion Splice Equipment is exceptionally user-friendly.

(A.) There are full sized core alignment instruments. (B.) Others are high performance v-groove cladding alignment devices. (C.) Others are explosion proof units for use in a wide range of ambient conditions where the plasma arc could ignite latent gasses. (D.) Consider "mini" designs that can be used on site in 'all axis' inverted, positions. (E) Ribbon fiber v-groove alignment fusion splicers represent past and future deployments.

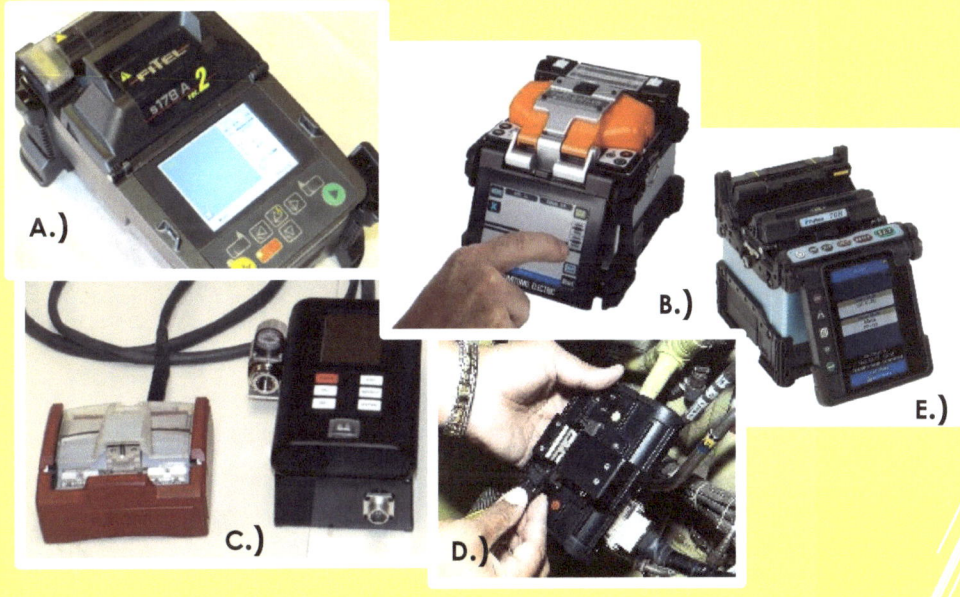

A.)

B.)

C.)

D.)

E.)

Compare and contrast: which advantages do you need. Look to the future to protect your business interests.

Always maintain this sophisticated equipment with highest quality precision cleaning materials and techniques to protect the investment.

Did You Know?
Many current fusion splicers can be rented?
"Try before you buy"!

These rugged instruments not only provide high levels of low loss, first-time splicing, but also they "talk to the user" and "trouble-shoot".

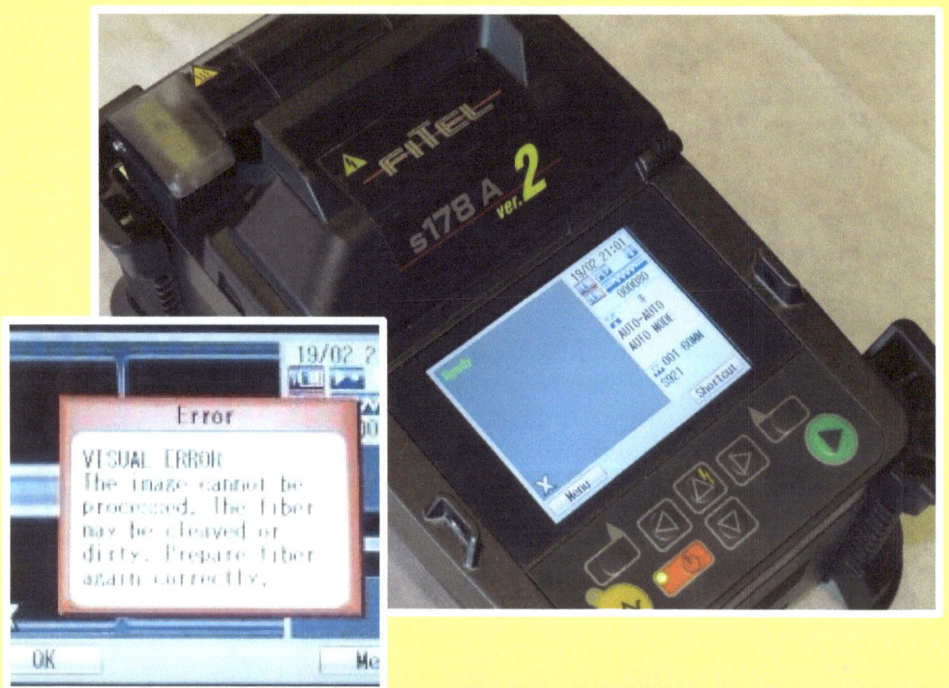

VISUAL ERROR:
"The image cannot be processed. The fiber may be cleaved or dirty. Prepare fiber again correctly."

In this actual splice, 1 fiber was clean and the other intentionally not cleaned to demonstrate this feature.

The newest equipment is self-diagnostic.
What the equipment does not and cannot discern is the type of debris or contamination. How to precision clean it is your responsibility. This work is created to help with those decisions.

Always use
"critical cleaning" products

WHY NOT COTTON?

Cotton is an ideal material for "thread" and here you see cotton "threading". **1** This linting material can cross contaminate lenses-to-mirrors, v-grooves to electrodes...anyplace to anywhere!

WHY NOT PAPER?

Paper is absorbent, but it does not have tensile strength. **2** When it tears it shreds. When it absorbs too much liquid, it disintegrates. **3**

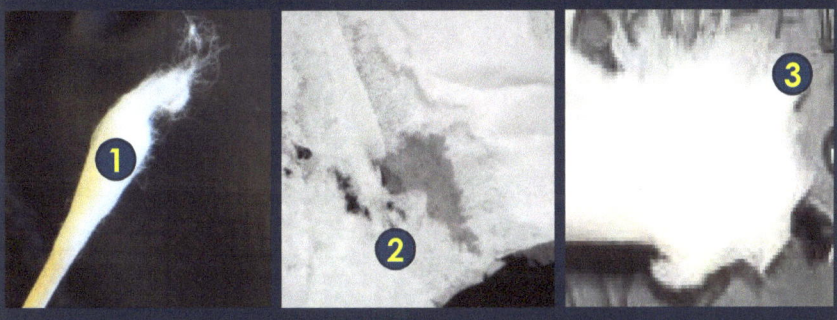

Don't take a chance! Update and upgrade to precision cleaning materials.

Selecting Cleaners and Wiping Materials

- How well does the solvent clean?
- What's expected a wiping cloth or swab tool?
- How is it packaged? Is Shipping important?
- What's the E H & S (environmental, health and safety) impact?

SELECTING A FIBER OPTIC CLEANER IS NOT ALWAYS STRAIGHT FORWARD AS ANSWERING "YES" OR "NO"!

The producer of a product is required to create a Material Safety Data Sheet, or MSDS. Be sure you understand how to read the MSDS: *simply having one doesn't assure a product is "safe" or "effective".*

After worker safety-first: the concerns are: 1.) What is the debris?   2.) Is this product actually capable of cleaning this debris? 3.) Is there possible damage to (plastic) components?   4.) In doubt? Request a sample or ask the producer a TDS and 'references'!

A Technical Data Sheet or TDS will give you important and specific information about such things as "plastic compatibility, flammability and other product attributes. The TDS is important: challenge your supplier by asking "good questions": don't accept a sales story!

SELECTING A WIPING MATERIAL IS A LITTLE EASIER.

Ask if the material could be used in a "rated" clean room, wafer fab, microelectronics or medical device production.

The response will give you an <u>indication</u> of the intrinsic 'cleanliness' of a specific wiping material. It's a start place…from there, test & compare!

What's Wrong with IPA?

There are three critical aspects to use of IPA...and this includes 99.9% 'reagent grade'... the good stuff!

1.) <u>Isopropyl alcohol is a disinfectant and good cleaning agent for fingerprints: it removes chewing gum and tree sap,</u>

However, even in the ~95-99.9% range (as often used for this application) isopropyl alcohol is not an effective cleaner on many type contaminants found in fiber optic deployment

The question is: "How do you know?".
The answer is: "It's not realistic to guess!".

2.) <u>Isopropyl alcohol is hygroscopic:</u> this means it attracts moisture to itself. A 2003 study of 99.9% IPA showed that a measured container absorbed >3% moisture in less than fifteen minutes. "Retail-store IPA" may be 70-90% IPA with 30-10% moisture. ☺
("I was out of IPA so I went to the drug store" ☺)

3.) <u>A major factor is storage:</u> IPA is most-typically stored in "squeeze-type" or "pump(Menda®) containers". As the solvent is squeezed out, moisture is brought into the container.
Worse yet: a "host container" used to fill a pump bottle has <u>head room</u>" that adds moisture and amplifies the problem.

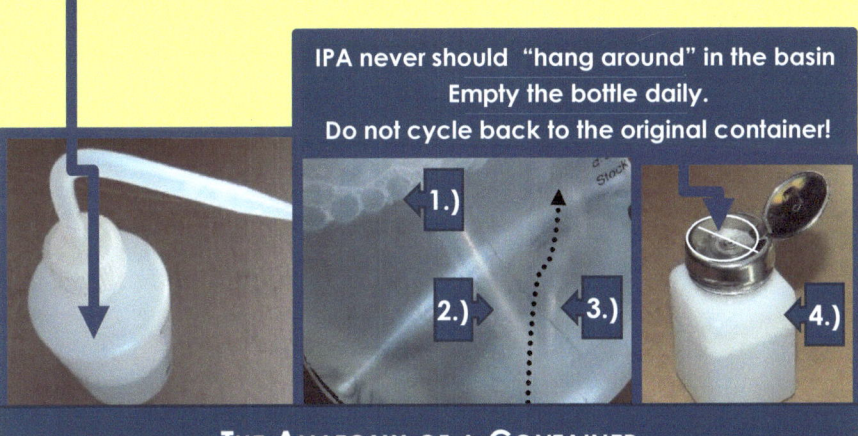

IPA never should "hang around" in the basin
Empty the bottle daily.
Do not cycle back to the original container!

1.)   2.)   3.)   4.)

THE ANATOMY OF A CONTAINER

As solvent is 'squeezed out", air is brought in as: 1.) the fluid level in the container (that was solvent) is replaced through 2.) the delivery tube, 3.) as air contaminates (note bubbles) the remaining fluid level and 4.) creates ambient air "head room".

How to select a fiber optic cleaner: Worker safety and environmental concerns are always top criteria. Be sure you read and have MSDS explained to you: ask the provider. Merely *having* an MSDS is not the point OSHA intended to educate you...it's important to understand the document itself.

This range of chemicals represents the 2015 list of best choices. These are general chemical families and each company has a 'trade name' for their product. Ask your supplier *"what's in this stuff?"*.

To "future-proof" you also have to self-educate.

HFE 7100 w/IPA/CZ®	Precision Hydrocarbon Formulations	Aqueous (Glycol Esters)
▸ Advantages	▸ Advantages	▸ Advantages
▸ Numerous formulations	▸ Numerous formulations	▸ Newest formulations
▸ Very good cleaning	▸ Wide range cleaning	▸ Wide range cleaning
▸ Check	▸ Check	▸ Check
▸ Convenience containers	▸ Convenience Containers	▸ Growing demand in many segments
▸ Easy Ship	▸ Low cost	▸ Convenience Containers
▸ Aerosol	▸ Disadvantages	▸ Easy Ship
▸ Non-flammable	▸ DOT regulated shipping	▸ Lowest cost
▸ Disadvantages	▸ As with IPA	▸ Disadvantages
▸ Ultra-Fast Evaporating		▸ Must dry with a 'wet-to-dry" step
▸ Can leave residues		
▸ Highest cost		

Test and compare. Try not to let 'aroma' be a criteria unless there are specific and individual reactions. *A chemical choice is based on many factors and* (followed by safety matters) *the actual cleaning ability is the foremost consideration*. Demand training from your supplier: it may be no charge or fee based.

How to select a fiber optic wiping material

Many types of cloth have been created since the beginning of time. Earliest came from animal skins which were superseded by weaving. Now, there are synthetic animal skins and woven cloth of natural and synthetic materials.

There is also a new-generation of material that weaves or hydro-entangles synthetic and natural materials. Selection of a strong material that does not shred or leave a surface residue is a critical concern to the precision fiber optic cleaning process.

100% Cellulose (paper)	Polyester / Microfiber	Hydroentangled polyester/cellulose

▸ Advantages
 	
  ▸ Absorptive
 	
  ▸ Convenience package
 	
  ▸ Readily Available
 	
  ▸ Lowest Cost
 	
▸ Disadvantages
 	
  ▸ Low sheer strength
 	
    ▸ Tears, shreds with Lint residues
 	
  ▸ Used for many applications
 	
    ▸ Embedded in supply chain
	▸ Advantages
	
	  ▸ Absorptive on most debris
	
	  ▸ High sheer strength
	
	    ▸ Low Linting
	
	  ▸ Often used in probes and other cleaning devices
	
	    ▸ Readily available
	
	  ▸ Moderate cost
	
	▸ Disadvantages
	
	  ▸ Can create a static charge
		▸ Advantages
		
		  ▸ Highly absorptive
		
		  ▸ High sheer strength
		
		  ▸ Often used in cleaning platforms
		
		    ▸ Readily available
		
		  ▸ Moderate cost
		
		▸ Disadvantages
		
		  ▸ Many choices and not all perform to the same level.

Test and compare. 100% cellulose (paper) and 100% cotton are not recommended to precision clean a fiber optic connection. Likewise, any treated material (possibly with an electro-static discharge [ESD] compound) is not suggested as there can be a residue.

Selection of Wiping Materials

Lens-grade and hydroentangled non-woven cellulose/polyester blends are some of the most cost effective and high performance wipers available.

Lens grade tissues are acceptable to clean the optical surfaces of fusion splicer but should not be used for preparation of the glass fibers. They may have unacceptable chemicals that deposit residues and foul electrodes.

1 This cellulose/polyester wiper has exceptional strength. This is an indication it will not leave a lint residue.

2 In this image, the wiper is distorted...but does not shred or tear. *Now consider: does it absorb?*

Perform simple "tear-tests"... if the wiper tears easily like a Kleenex® or KimWipe® it is *not acceptable* to precision clean any fiber optic surface.

Micro-fibers are good...but there are many levels. Some are used for clothing while others are "cleanroom grade"!

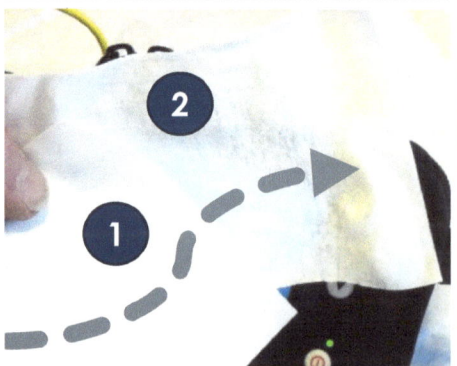

There may be as many as 4,000 different cellulose/polyester combinations. Some have too much polyester and are not absorbent and others lack strength.

There are superior 100% polyester wipers: they are not as absorbent as "cleanroom microfibers" and cellulose/polyester! 100% polyester can create a static field that attracts additional dry debris!

Challenge your supplier. Request samples.

Don't use "too much" of anything!

a. *Never 'direct-spray'*

b. Spray into the wiper

a.

b.

b. 'Transfer clean' by using a lint-free hyrdoentangled wiper.

c.) Move the wiper in one direction, left to right, for example. Do not re-trace by moving 'back and forth' or in a circular motion.

c.

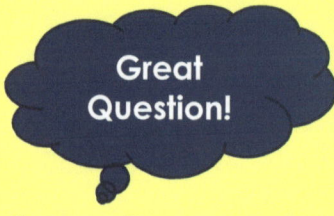

How do you know it's clean?

Look at the cleaning material. *Have you ever wondered why high-quality swab tools and wipers are white?*

It's because you will be able to see debris on the white background.

Always look at the wiper or swab to assure you have removed the soil!

Simple observations really work!

Cross-contamination occurs when a contaminated wiper or precision swab tool is re-used. The best practice rule of them is to use precision cleaning materials one time.

Cleaning "Do and Do Not"

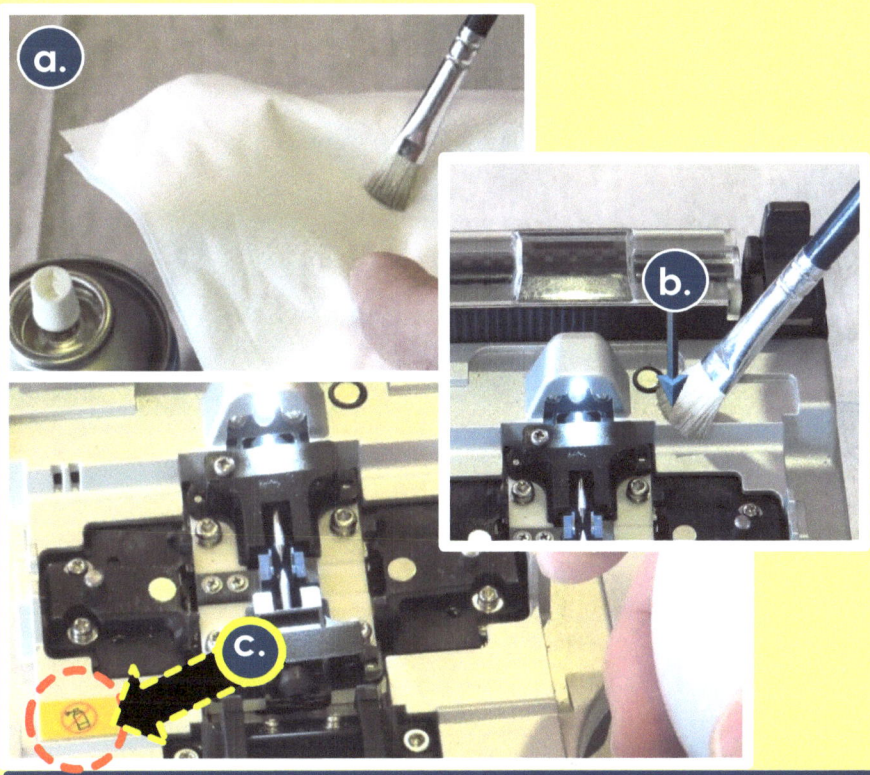

a.) Under-use all cleaners by transferring the cleaner from a wiper to the cleaning tool.

b.) Notice the dark color of the brush indicating accumulation of debris. This tool is doing its' job...it will need cleaned or replacement at some point.

c.) Many producers caution against using compressed gas dusters or compressed shop air. Safety note: 1.) there may be glass shards that can be launched by these 'blasts". 2.) Some compressed gas dusters have 'flammable' gas that can be ignited by the fusion splice arc. *If you must...always wear safety glasses and assure the splicer has no residue: use compressed dusters with 'all-way' valves and non-flammable components. Read MSDS carefully to assure "non-flammable" really is!*

Some compressed gas dusters are just not effective on debris and contamination no matter how strong they may 'huff and puff'. In reality, "Precision Cleaning" often means 'touching and removing' specific debris or contamination.

PRECISION CLEANING

Each manufacturer produces a slightly different machine. Some components are serviceable in the field. Other machines may have components that require service at a factory depot.

Check your manual and ask questions before making a selection 'which to buy'.

How is are fusion splice machine components precision cleaned?

Your fusion splicer is a technical marvel. It is a computer controlled welder that will successfully fuse single and multiple fibers from ribbon cable.

Here is what must be precision cleaned to assure best results:

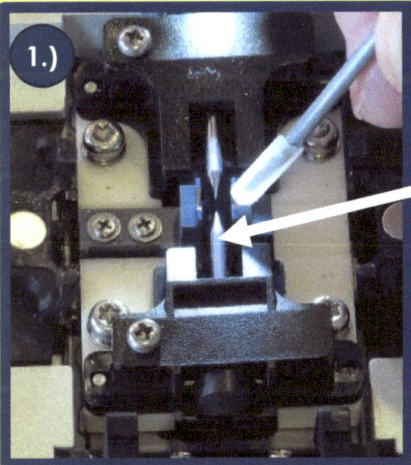

1.) Precision Clean the electrodes. These are the "tips" and the business end of the device.

They may be soiled and they also may corrode as a part of the fusion spice process as the arc burns off impurities or moisture...perhaps from hygroscopic IPA.

2.) Never oversaturate a precision swab by 'dunking it' in a pool of any cleaner.

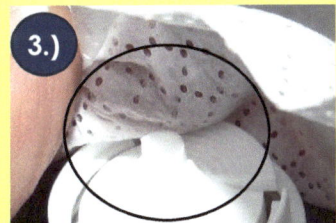

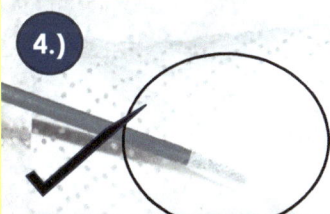

3.) Lightly moisten the wiper. 4.) Use the solvent transfer technique to saturate the swab head for a count: **1-2-3-4-5.** (about 3-5 seconds!)

Some producers of fusion splice equipment provide means to clean the electrodes. Others require replacement to restore the precision surface. Electrodes can become contaminated or 'corroded' as a normal process of the welding procedure. Some theorize that moisture from hygroscopic IPA causes this corrosion.

Consider high-performance fusion splice preparation cleaners as alterative to 99.9% IPA.

This is a micro-particle embedded 'eraser-like' material that micro-polishes the electrodes.

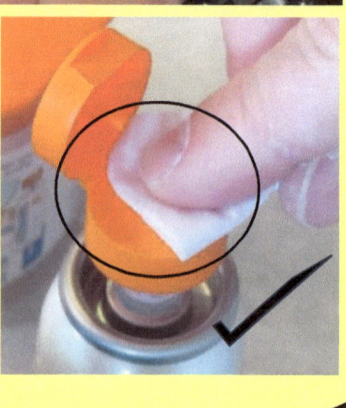

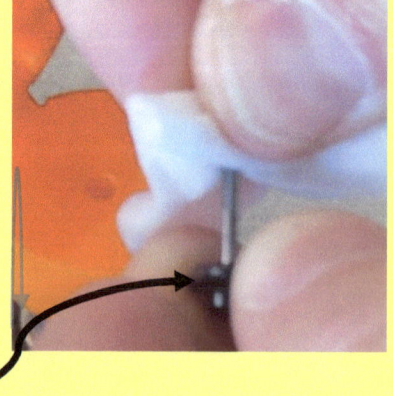

After micro-polishing the electrode must be cleaned. *Remember, always use less of any chemical any time cleaning anything related to fiber optics.*

Cleaning Key Fusion Splicer Components
(Check your manual to see which of these are present and if they are field-serviceable.)

1.) <u>V-Grooves.</u>

V-Grooves are critical components to every fusion splicing unit. Be sure you are aware where they are located on your individual device.

The V-Groove situates a single fiber or multiple fibers for the mass splice for the x-y-z axis alignment.

1.) If V-Grooves are not clean, then there is potential the fusion splicer's computer can misalign the splice. Lightly pass the swab tip through each groove. Never use a metal brush. A cotton tip is as troublesome.

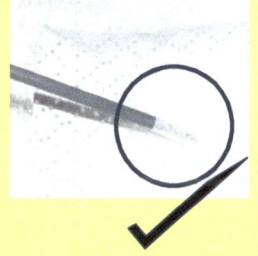

2.) Don't use too much of anything! Use the transfer technique to moisten swab tips!

21

Cleaning Key Fusion Splicer Components

2.) <u>Lenses.</u>

On other equipment styles, the lenses are easily accessed.

Since every machine is slightly different, read the manual to identify the location of these components.

Select the proper size swab tool that "fits"… *never "force fit" a cleaning tool into a small precision area.*

Some lenses may be be coated. For this reason I recommend when moistening a swab tip, simply use "purified water" and always 'solvent transfer'.

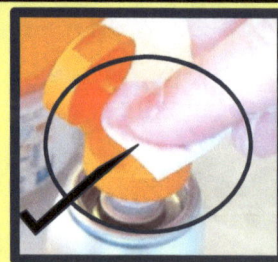

<u>Applications Tip:</u>

What's "purified water"?   Many brands possible. Why not "tap water"…no reason "why not" except for possible mineral content.   "Purified water" is likely on your work bench right now!

Cleaning Key Fusion Splicer Components

3.) <u>Mirrors</u>.

Often overlooked component of fusion splicer are the mirrors present on some devices.
Each component of the fusion splicer acts in harmony to deliver a low loss first time splice.

Check your owner's manual to determine the lens location and how to access them.

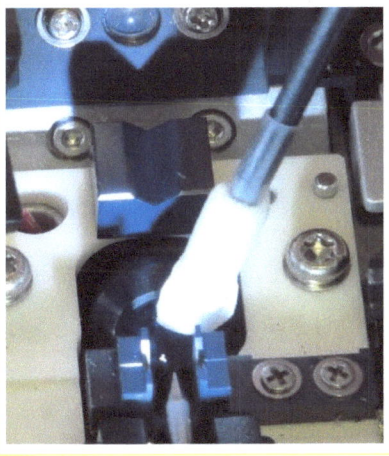

1.) As with lenses, mirrors may be cleaned with 'cleanroom puffs' and others with different swab tools.

2.) Some may be coated. For this reason I recommend when moistening a swab tip, simply use "purified water" and always 'solvent transfer'.

3.) Minor disassembly possibly required on some devices to access lenses.

Cleaning Key Fusion Splicer Components

4.) LEDs.

Each component of the fusion splicer acts in harmony to deliver a low loss first time splice. If your equipment has been in "severe duty", it's best practice to clean each individual component.

Check your owner's manual to determine the location of all components and whether or not you can service them.

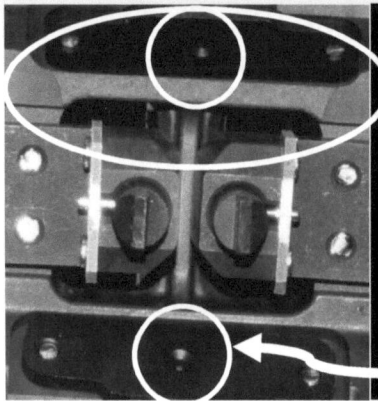

The smaller the component the more sensitive it can be to contamination and performance loss.

If the area surrounding a component is contaminated with dust, then it's a good indication a major component such as an LED may also be.

5.) Fiber Holders

I always like to clean with a lightly moistened swab tool.

When precision cleaning, think as though you would remove contamination from this 'super car'.

No scratches, no static-charge dust accumulation, no regrets

The Four Steps to Fusion Spice

CLEAN, STRIP AND CLEAVE THE FIBER

1.) Clean the Fiber
(Single or Ribbon)

The term "fiber" itself means different things at different points to different persons!

In this instance, gel from old style fiber ribbon is cleaned with a pre-saturated non-IPA wiper before the outer matrix is removed for mass fusion splice.

Here the "glass" or fiber is cleaned prior to being placed in a fiber holder and into the fusion splice machine. Some fibers may require cleaning 2-3 times to remove blocking gels or other residues from the 'clean and strip'. Some fibers will 'squeak' clean!

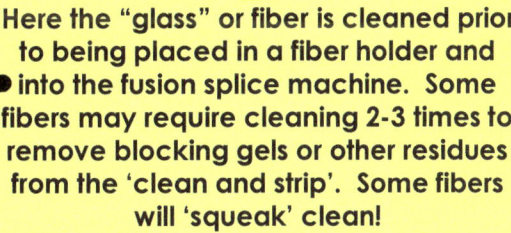

Proper procedure is critical to low loss splices.

There can be many contaminants on ribbon matrix. Many in the industry have started using precision hydrocarbons: they remove most contamination types.

Be wary of terpenes (smell like citrus). These can leave residues which themselves, in turn, must be cleaned.

"A tip from an Old Splicer"
"The best alternative, especially for high strength and pm splicing. Is dipping the fiber in an ultrasonic cleaner"

Or, use one of the "High-Po" cleaners listed on Page 12 for field work

2.) Strip the Fiber

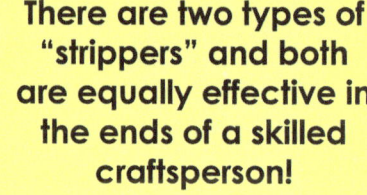

There are two types of "strippers" and both are equally effective in the ends of a skilled craftsperson!

Mechanical strippers are high precision and relatively inexpensive means of removing outer coatings. These are used for single fibers only.

Hot strippers are used for both single fibers and are highly effective to strip ribbon fiber.

There are mechanical as well as also a patented chemical mid-span break-in processes for ribbon fiber.

The Four Steps to Fusion Spice

3.) Cleave the Fiber

Lightly moistened swab tool here!

CLEAN THE PRECISION SURFACES OF THE CLEAVER !

- Assure cutting wheels are residue free as well as "Guillotine" surfaces.
- Look carefully: if any surface is dusty or has a residue: clean it!

Lightly moistened swab tool here!

The cleaver is a precise guillotine-like device that prepares the fiber end for the actual splice.

This is a precision device that likely has a specific number of cuts before the blades are replaced.

Some producers have cleavers with automatic rotating blades. This means the same blade surface is not being used sequentially.

Some cleavers are designed for both single and ribbon fibers.

Since each tool is slightly different, actual cleaning will vary! Check the instructions.
Ask the manufacturer!

Many splicers say their cleaver is one of the most critical tools in the tool-sack!

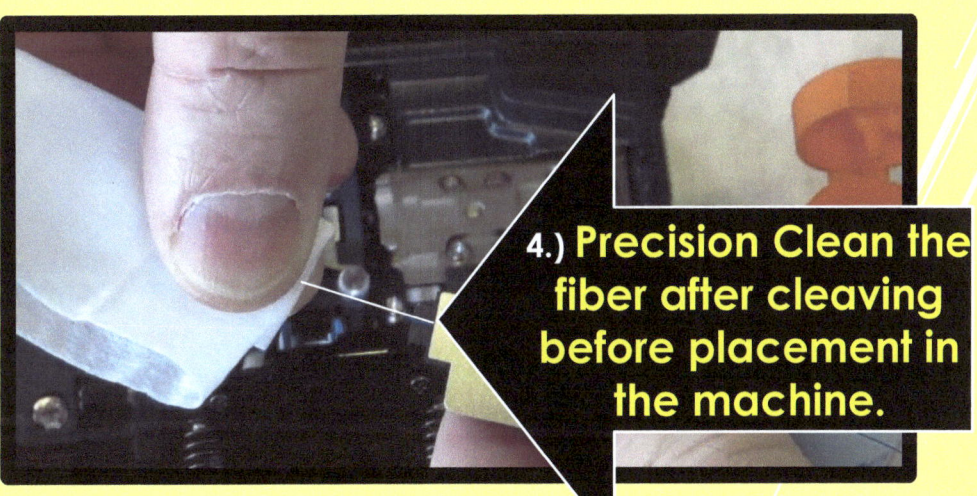

4.) Precision Clean the fiber after cleaving before placement in the machine.

27

Splicing

After stripping and cleaving, the fiber is placed in the fusion splice machine.

Since there is a variety of machine types, the 'bare fiber; is either placed in either fiber holders or removable fiber holders. All components must be cleaned to assure accurate alignment of the fusion splice.

1-Ready

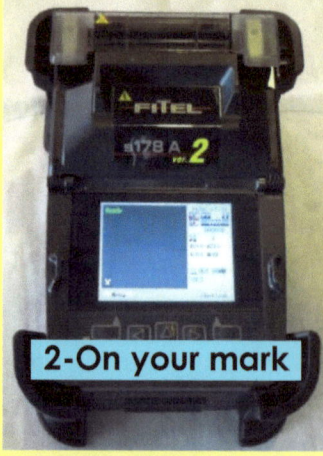

2-On your mark

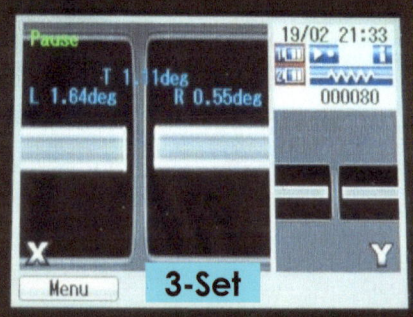

3-Set

4-Go

When every aspect of the splice process is carefully considered, the result is a low-loss first time event.

Cosmetic Cleaning

Yes, it's important that the equipment "looks clean" because surfaces can be cross-contaminated

What should be cleaned?

Cosmetic Cleaning compared to Precision Cleaning

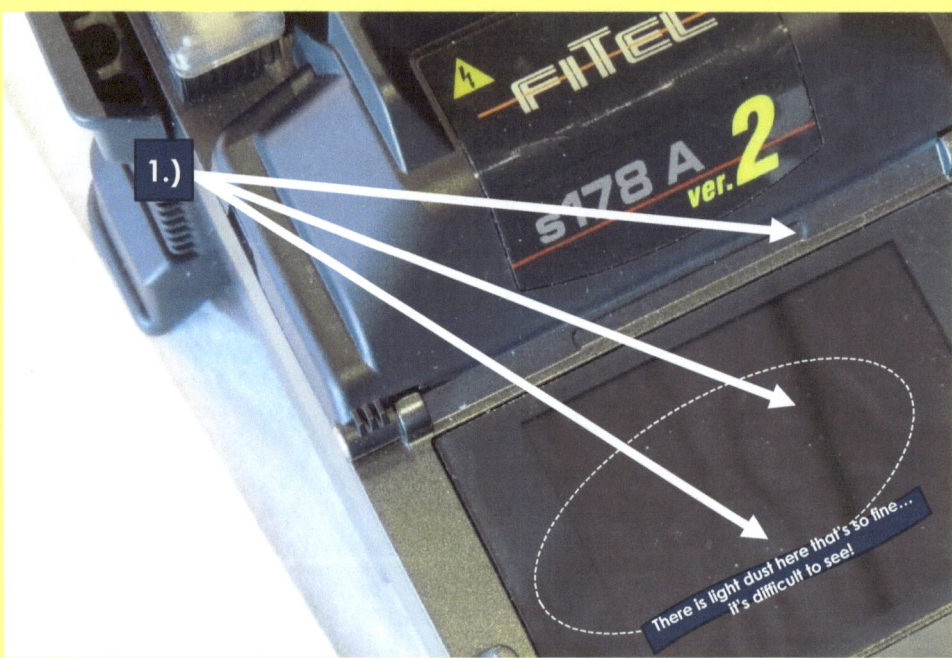

1.) There is light dust here that's so fine...
It's difficult to see!

1.) It is a "good assumption" that any dust of the chassis of a fusion splicer will end up on the internal mechanisms ...*even the lightest type you can barely see in this picture...on* the outside.

Some light dust is merely an annoyance and some may contaminate vital operational components. Therefore, treat all "cosmetic" cleaning as critical cleaning.

As rugged as it is, don't permit your expensive equipment to deteriorate.

Why is "cosmetic cleaning" essential to an effective fusion splice?

There are many components to a successful fiber optic fusion splice. Whether you are new to this you may be thinking "this is a lot to remember". If you have done a few thousand splices, these procedures are likely 2nd nature to you.

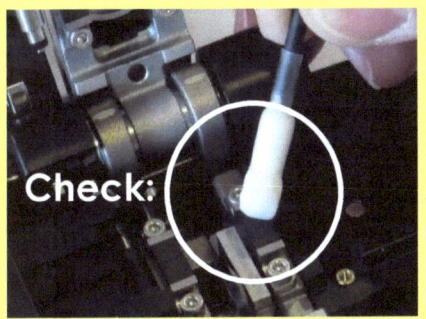

Check:

When was the last time you cleaned the cleaver?

Check:

When was the last time you cleaned the fiber clamps?

Check:

When was the last time you replaced the cleaning brush?

These tools to not last forever!

Creating an activity log for important equipment establishes a record that helps you decide if the equipment met your expectations. The log may also be a good tool when you want to sell it or trade it in!

Unit ID	Date	Activity	Technician
1	3.3.14	Cleaned V-Grooves	TD
1	4.1.14	Replaced electrodes at 2,300 cycles	TD
1	10.2.14	Returned to OEM.	TD
1	12.1.2014	Cleaned V-Grooves, Mirrors, Lenses after Patrick Energy job.	TD
1	12.31.14	Sold to 1-2-3 Ind.	

Conclusions

The fiber optic medium is an ongoing deployment of new technologies.

- In 2015, transmissions in the gigabit ranges are common; in July-2014 a new world record of more than 40 terabits was set over a single fiber.

- This means that we, as researchers, designers and craftspersons must understand what some might called minor or nuanced matters: *in this case assuring all aspects of a fiber optic splice are considered.*

- While this work discusses fusion splice, other aspects such as end face cleaning and video inspection are integral to the loss budget of any specific installation.

- If the connection is housed in a clean environment, perhaps once a month is a sufficient cleaning interval. If the connection is made on a feed lot in Western Kansas, military theatre in a desert region, or, entertainment center along a sandy beach...the PM might be daily. It is the environment and ever increasing demands and expectations that mandates increased awareness to assure the tools-of-the-trade are matched to the demands-of-the-job.

- Assuring fusion splice equipment is precision-clean helps ensure long life of expensive investments.

Test Your Knowledge:

True	False	
		Core alignment machines key on the cladding which is 125 microns
		Core alignment machines key on the core which is 9 microns
		Even a skilled technician cannot make a low loss splice on a cladding alignment machine
		As a back up, fusion splice machines have manual settings for the x-y-z axis alignment
		Cosmetic cleaning of fusion splice machines is important because what is on the outside can get into the working parts of the instrument.
		An "explosion proof" fusion splicer would be valuable in mining, underground or any application where the plasma arc could ignite certain gasses.
		Cotton and paper products are as good as non-woven materials to precision clean fusion splice operational components.
		IPA is no longer acceptable because of its' high cost.
		Mirrors, lenses and V-Grooves are some of the precision components that must be maintained on all fusion splicers
		Fusion Splice loss is part of the overall loss budget that includes a 'fiber loss', 'connector loss' and 'splice loss'.
		Brushes and wipers can be used indefinitely.

Test Your Knowledge:

True	False	
	X	Core alignment machines key on the core or fiber or glass…which are inter-used terms.
X		Core alignment machines key on the core which is 9 microns
	X	A skilled technician can make a low loss splice on a clean and capable machine
	X	These instruments are fully automatic and highly capable devices. Keep them precision cleaned for best results.
X		If there is debris on the outside…it can permeate the inside.
X		An "explosion proof" fusion splicer would be valuable in mining, underground or any application where the plasma arc could ignite certain gasses.
	X	Cotton and paper products are acceptable to clean cosmetics…they are not acceptable precision cleaning materials.
	X	IPA is no longer acceptable because it does not clean a wide range of debris and because it is often stored in containers that weaken this low cleaning ability further.
X		Mirrors, lenses and V-Grooves are some of the precision components that must be maintained on all fusion splicers
X		Fusion Splice loss is part of the overall loss budget that includes a 'fiber loss', 'connector loss' and 'splice loss'.
	X	As you would a toothbrush or personal tissue, change these regularly!

ABOUT THE AUTHOR:

▸ Ed Forrest has been actively involved in specification and applications engineering of various precision cleaning applications for more than 25 years.   Previously employed at ITW Chemtronics®, retired in July-2014, he was schooled to analyze precision and gross cleaning applications in a wide range of applications. In 2001 he began development of a program that resulted in formal approvals at all major telecommunications providers.

▸ He has seven patents in the areas of fiber optic precision cleaning with six products in production and marketing credits that include branding, training, and publication of materials. He invented a chemical mid-span break-in for ribbon fiber, a widely-used cleaning tool, and identified precision cleaning products and a fusion splice prep delivery system. He has other patents pending.

▸ He is active on fiber optic standards committees and is considered a SME in the study of fiber optic cleaning and inspection. His work is based on field experiences and the needs of designers, crafts persons and production line workers.

▸ His practical thesis of "Five Zone Cleaning" is a look forward to the times when high speed and capacity of fiber optic transmission (even more) will be impacted by a contaminated or improperly cleaned connections. He has uniquely researched inspection of the 4th and 5th Zone and the influences of various debris and contamination as it is positioned on these areas of the connector.

▸ He worked as an Electronics Manufacturer's Representative throughout the 1970's. He actively participated in the early introduction of some of the most fundamental electronic products in the changeover from analogue to solid state. These included solid state components, consumer products including the first hand-held calculators, esoteric high fidelity, test equipment, games and other electronic products considered 'cornerstones' of the contemporary marketplace. He has production credits in that Industry

▸ He worked in a then-developing market segment in the Home Furnishings Industry. By coordinating North American and International Development, using an effective agency in Denmark he was able to work throughout Europe prior to the time of the EU. In coordination with C.ITOH (est-1860) , he traveled and developed a Japanese market long before current interest in the important nations of The Pacific Rim. He initiated promotional activity in conjunction with USA Embassies, individual USA states resulting in active trade in Denmark, Sweden, Finland, Italy, Germany. Great Britain, nations in The Middle East and South Africa. He has production credits in that industry.

▸ Early career as a Technical Representative, in Union Carbide Corporation's Automotive Consumer Products group, career-forming experiences include introduction of Prestone® AntiFreeze as a Summer Coolant in a one year NASCAR race test and associated promotions, as well as, an innovative time with Standard Oil of Ohio® as SOHIO® introduced "self-service fueling" to the market. He competed in the market when brands like STP® and Wynn's® dominated consumer interest.

He is an active photographer, enjoys study of the ancients, and is a hobbyist collector of esoteric high-fidelity. As a life-long SCCA member he competed "wheel-to-wheel" in more than 200 events at SCCA's Club Racing level in cars he designed...'with a little help from his friends'. Married with a fascination for Weimaraners, he and his wife are often at the edge with three lovely specimens. They have travelled extensively throughout the USA and Europe.

Contact Information:
+770-971-8100 USA
edforrest@live.com
www.fiberopticprecisioncleaning.com

Want to Study More?

- Discussion of standards and why they are obsolete when they are published

- How to Create Internal Standards for your organization and why.

- "Science of Cleaning"

- How to clean a fiber optic connection when you don't have a video scope

68 Pages in full color. 6"x9" Paperback
www.createspace.com/ 5173068
www.amazon.com

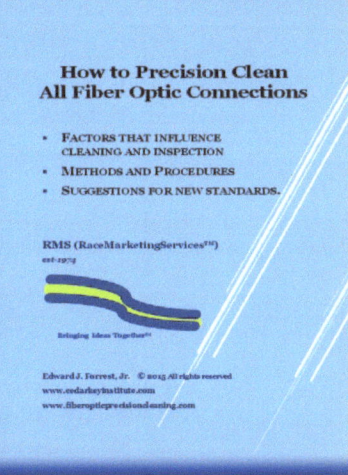

How to Precision Clean All Fiber Optic Connections

- FACTORS THAT INFLUENCE CLEANING AND INSPECTION
- METHODS AND PROCEDURES
- SUGGESTIONS FOR NEW STANDARDS.

RMS (RaceMarketingServices™)
est-1974

Bringing Ideas Together™

Edward J. Forrest, Jr.   © 2015 All rights reserved
www.cedarkeyinstitute.com
www.fiberopticprecisioncleaning.com

A COMPARISON STUDY OF PRECISION CLEANING METHODS FOR ALL FIBER OPTIC CONNECTIONS

- COMPARATIVE EVALUATIONS OF CONTEMPORARY PRODUCTS, METHODS AND PROCEDURES.
- FEATURING A STUDY OF A 2.5MM SC JUMPER AND SFP-TYPE OPTICAL TRANSCEIVER CONNECTION.
- WHAT WORKS; WHAT DOES NOT AND WHY.
- WHY AND HOW TO EXCEED STANDARDS.

RMS (RaceMarketingServices™)
est-1974

Bringing Ideas Together™

Edward J. Forrest, Jr.

www.fiberopticprecisioncleaning.com
www.cedarkeyinstitute.com

© 2014 All Rights Reserved

- New 3rd Edition

- Answers the questions: what works best? CleTop®, IBC®, Swabs, QbE®, Sticklers®.

- Discusses Zone-5 Contamination and how to avoid it.

- Eight "nasty soils" you might find on the job!

138 Pages in full color. 8.5x11" Paperback

www.createspace.com/ 5409877

www.amazon.com

- The significance of cleaning test equipment and how!

- What works and what to avoid

25 pages in full color. 6"x9" Paperback

www.createspace.com/52963 45

www.amazon.com

Understanding Cross-Contamination Points on Fiber Optic Inspection and Test Equipment

- HOW DIRTY EQUIPMENT CAN NEGATIVELY INFLUENCE VIDEO INSPECTION AND TEST/MEASUREMENT RESULTS

RMS (RaceMarketingServices™)
est-1974

Bringing Ideas Together™

Edward J. Forrest, Jr.   © 2015 All rights reserved
www.fiberopticprecisioncleaning.com

Updates and White Papers.
Please visit: www.fiberopticprecisioncleaning.com

Notes

DISCOVERY...
"The voyage of discovery is not in seeking new
landscapes but in having new eyes."
Marcel Proust (1871-1922)